The Hidden World: Exploring Forest Conservation Biology for Everyone

GALBRAITH

The Hidden World: Exploring Forest Conservation Biology for Everyone

Copyright © 2023 by GALBRAITH

The first edition was published in 2023

ISBN:
Published by:
Sunshine
1663 Liberty Drive
Hyderabad, IN 47403
www.Sunshinepublishers.com

This book is self-published using on-demand printing and publishing, which allows it to be printed and distributed globally

TABLE OF CONTENT

Chapter 6: Case Studies in Forest Conservation   50

Successful Forest Conservation Projects

Challenges and Lessons Learned from Forest Conservation Initiatives

Forest Conservation Efforts at the Global Scale

Local and Indigenous Perspectives on Forest Conservation

Chapter 7: The Role of Individuals inForest Conservation   58

Everyday Actions for Forest Conservation

Getting Involved in Forest Conservation Organizations

Advocacy and Policy Change for Forest Protection

The Power of Education and Awareness in Forest Conservation

Chapter 8: The Future of Forest Conservation   66

Chapter 9: Conclusion   74

Chapter 1: Introduction to Forest Conservation Biology

The Importance of Forest Conservation

Forests are not just a collection of trees; they are intricate ecosystems that support countless species and provide numerous benefits to humanity. In this subchapter, we will delve into the crucial role of forest conservation and why it is essential for everyone to understand and actively participate in preserving these natural wonders.

Forest conservation biology is a field that focuses on understanding the complex interactions within forests and the impact of human activities on their health. By studying forest ecosystems, scientists gain insights into the delicate balance of biodiversity, climate regulation, and resource sustainability that forests provide.

One of the primary reasons forest conservation is vital is its role in preserving biodiversity. Forests are home to an astounding array of plant and animal species, many of which are found nowhere else on Earth. By conserving forests, we protect these species and maintain the delicate web of life that depends on their existence. From colorful birds to elusive mammals and fascinating insects, each organism plays a unique role in the forest ecosystem.

Moreover, forests act as the lungs of our planet, absorbing carbon dioxide and releasing oxygen through photosynthesis. This process helps regulate the Earth's climate and mitigates the effects of global warming. Deforestation, on the other hand, releases vast amounts of stored carbon into the atmosphere, contributing to climate change. By conserving forests, we can actively combat climate change and its detrimental impacts.

Forest conservation also has significant socioeconomic benefits. Forests provide invaluable resources such as timber, fuelwood, medicinal plants, and food. Many communities around the world rely on forests for their livelihoods, and sustainable forest management ensures these resources are available for future generations. Additionally, forests offer recreational opportunities, attracting tourists and providing a source of income for local communities.

Understanding the importance of forest conservation is crucial for everyone, regardless of their background or profession. Whether you are a conservation biologist, a student, a policymaker, or simply a nature enthusiast, recognizing the value of forests and their conservation is essential.

By learning about forest conservation biology, we can make informed decisions about our actions and contribute to the preservation of these precious ecosystems. Through sustainable practices, reforestation efforts, and raising awareness, we can protect forests for future generations, ensuring the continued provision of vital ecosystem services and sustaining the incredible biodiversity that makes our planet thrive.

In conclusion, forest conservation is not just an issue for conservation biologists; it is a matter that concerns us all. Understanding the importance of forests and actively participating in their preservation is crucial for the well-being of both humans and the natural world. Together, we can create a future where forests continue to thrive, providing us with their invaluable benefits for generations to come.

The Role of Forests in the Ecosystem

Forests are not only enchanting and awe-inspiring, but they also play a crucial role in maintaining the delicate balance of our planet's ecosystems. In this subchapter, we will explore the profound significance of forests in the field of conservation biology and how they contribute to the well-being of all living beings.

Forests are often referred to as the lungs of the Earth, and for good reason. They absorb carbon dioxide, a greenhouse gas responsible for climate change, and release oxygen, which is essential for all life forms. This process, known as photosynthesis, helps regulate the global climate and maintains a stable atmosphere. Forests act as a natural carbon sink, mitigating the impacts of human activities such as burning fossil fuels and deforestation.

Moreover, forests provide habitat and shelter for an astonishing array of plant and animal species. They are biodiversity hotspots, home to countless organisms that rely on the complex web of interactions within the forest ecosystem. From majestic, towering trees to tiny, elusive insects, each species plays a vital role in maintaining the ecosystem's balance. Forests harbor immense genetic diversity, which is crucial for the survival and adaptation of species to changing environmental conditions.

Forests also serve as natural water filters, protecting watersheds and regulating the flow of water. Their dense canopy prevents soil erosion, ensuring that rivers and lakes receive clean, clear water. Additionally, forests act as natural sponges, absorbing rainwater and releasing it gradually, preventing floods and maintaining a stable water supply for humans and wildlife alike.

Another critical function of forests is providing livelihoods for millions of people worldwide. Forests are a source of food, medicine, and raw materials for countless communities, especially in developing countries. Sustainable forest management practices are necessary to ensure the continued availability of these resources while preserving the integrity of the ecosystem.

However, forests are facing unprecedented threats due to deforestation, illegal logging, climate change, and habitat fragmentation. These challenges necessitate urgent action to protect and restore forest ecosystems. Conservation biologists play a pivotal role in studying and understanding the intricate ecological processes within forests, devising innovative conservation strategies, and promoting sustainable practices.

In conclusion, forests are not just an enchanting backdrop to our lives, but they are indispensable for the health and survival of our planet and all its inhabitants. Understanding the role of forests in the ecosystem is crucial for everyone, regardless of their background or profession. By appreciating the significance of forests and supporting conservation efforts, we can ensure the preservation of these magnificent ecosystems for generations to come.

The Threats to Forests

Forests are invaluable ecosystems that play a crucial role in maintaining the health of our planet. They provide us with clean air to breathe, water to drink, and are home to countless species of plants and animals. However, forests are facing numerous threats that pose a significant risk to their survival. In this subchapter, we will delve into the various challenges that forests are currently facing.

One of the most pressing threats to forests is deforestation. This human-driven activity involves the clearing and removal of trees, often to make way for agriculture, urban development, or commercial logging. Deforestation not only disrupts the delicate balance of the forest ecosystem but also contributes to climate change. Trees act as carbon sinks, absorbing and storing carbon dioxide, a greenhouse gas that is a major driver of global warming. When forests are destroyed, this stored carbon is released back into the atmosphere, exacerbating the problem.

Another significant threat to forests is habitat fragmentation. As human populations continue to grow, forests are being divided into smaller and isolated patches by roads, agriculture, and infrastructure development. This fragmentation disrupts the movement of species, making it harder for them to find food, mates, and suitable habitats. It also increases the risk of inbreeding and reduces genetic diversity, which weakens the overall resilience of forest ecosystems.

Climate change is yet another major threat to forests. Rising temperatures, changing precipitation patterns, and extreme weather events such as droughts and wildfires are all impacting forest health. These changes can lead to increased tree mortality, reduced growth rates, and shifts in species composition. Forests are also vulnerable to invasive species, which can outcompete native plants and animals, leading to a loss of biodiversity.

Furthermore, unsustainable logging practices pose a significant threat to forests. Illegal logging, clear-cutting, and selective logging can all have detrimental effects on forest ecosystems. When trees are removed without proper planning or consideration for the long-term health of the forest, it can lead to soil erosion, loss of wildlife habitat, and disruption of natural regeneration processes.

It is crucial that we address these threats to forests and take immediate action to protect and conserve them. This requires a multi-faceted approach involving governments, organizations, and individuals. Conservation biology plays a vital role in understanding the impacts of these threats and developing strategies to mitigate them. By raising awareness, promoting sustainable practices, and supporting conservation efforts, we can ensure the preservation of our forests for future generations. It is up to all of us, regardless of our backgrounds or areas of expertise, to come together and work towards a sustainable future where forests thrive and continue to provide us with their invaluable services.

The Need for Forest Conservation Biology

Forests are not only mesmerizing and awe-inspiring but also play a crucial role in sustaining life on Earth. They are home to an astonishing array of plants, animals, and microorganisms that collectively form an intricate web of life. However, in recent decades, forests have been facing unprecedented threats due to human activities such as deforestation, habitat destruction, and climate change. This subchapter aims to shed light on the urgent need for forest conservation biology, a multidisciplinary field that strives to understand and protect these invaluable ecosystems.

One of the primary reasons why forest conservation biology is crucial is the alarming rate of deforestation. Each year, vast areas of forests are cleared for agriculture, logging, and urbanization. This destruction not only results in the loss of biodiversity but also disrupts the delicate balance of the ecosystem. Forest conservation biology seeks to address this issue by studying the ecological impacts of deforestation and developing strategies to mitigate its effects.

Moreover, forests act as vital carbon sinks, absorbing and storing large amounts of carbon dioxide from the atmosphere. With the increasing levels of greenhouse gases contributing to climate change, the preservation of forests becomes even more critical. Forest conservation biology plays a pivotal role in understanding the role of forests in combating climate change and devising methods to enhance their carbon sequestration capabilities.

Furthermore, forests provide numerous ecosystem services essential for human well-being. They regulate water cycles, purify air and water, provide food and medicinal resources, and offer recreational opportunities. Forest conservation biology aims to safeguard these services by studying the relationships between forests and the surrounding environment, and by implementing sustainable management practices.

For everyone, forest conservation biology offers an opportunity to engage with nature and contribute to the preservation of these magnificent ecosystems. Whether you are a student, a scientist, a policy-maker, or simply someone passionate about the environment, understanding the importance of forest conservation biology is crucial. By raising awareness about the threats faced by forests and the need for their protection, we can collectively work towards sustainable solutions.

In conclusion, the need for forest conservation biology has never been more urgent. In the face of escalating deforestation rates, climate change, and the loss of biodiversity, this multidisciplinary field provides the knowledge and tools to address these challenges. By embracing forest conservation biology, we can ensure the long-term survival of forests and secure a sustainable future for all.

Overview of the Book

"The Hidden World: Exploring Forest Conservation Biology for Everyone" is a captivating and enlightening book that takes readers on a journey into the captivating realm of forest conservation biology. Written with the intention of being accessible to everyone, this book aims to educate and inspire individuals from all walks of life, regardless of their background or expertise, to understand and appreciate the importance of conserving our forests.

The book begins by providing a comprehensive introduction to the field of conservation biology, unraveling the key concepts and principles that underpin the study of forests and their conservation. It delves into the intricate web of life that exists within forests, highlighting the interdependencies between species, and the crucial role that forests play in maintaining the ecological balance of our planet.

Throughout the chapters, readers are introduced to a wide range of topics, including the impact of deforestation, climate change, and habitat loss on forest ecosystems. The book examines the various threats faced by forests and the biodiversity they support, shedding light on the consequences of these challenges for both wildlife and humanity.

"The Hidden World" also explores the innovative strategies and approaches employed by conservation biologists to protect and restore forest ecosystems. From community-based conservation initiatives to cutting-edge technology, readers will gain insights into the diverse range of methods utilized to promote sustainable forest management and safeguard biodiversity.

Moreover, the book emphasizes the significance of engaging every individual in the conservation efforts. It encourages readers to take an active role in preserving forests, offering practical tips and suggestions for incorporating sustainable practices into their daily lives.

Drawing on real-life examples and case studies from around the world, "The Hidden World" showcases the remarkable beauty and complexity of forest ecosystems. Through vivid descriptions and stunning visuals, readers will be transported into the heart of the forest, fostering a deep sense of connection and appreciation for these incredible natural wonders.

In conclusion, "The Hidden World: Exploring Forest Conservation Biology for Everyone" is an accessible and thought-provoking book that aims to ignite a passion for forest conservation in readers from all backgrounds. Whether you are a student of conservation biology or simply an individual interested in understanding and protecting our forests, this book will empower you to make a positive impact on the hidden world that lies within our forests.

Chapter 2: Understanding Forest Ecosystems

Forest Types and their Characteristics

Forests are incredibly diverse ecosystems that cover vast areas of our planet. They provide essential habitats for countless species, regulate climate patterns, and offer numerous resources for human communities. In this subchapter, we will delve into the different forest types found around the world and explore their unique characteristics.

1.   Tropical   Rainforests: Tropical rainforests are characterized by high levels of rainfall and humidity throughout the year. They are home to a wide variety of plant and animal species, many of which are endemic. The dense canopy of these forests creates a complex and interconnected ecosystem, fostering incredible biodiversity.

2.   Temperate   Forests: Temperate forests are found in regions with moderate climates and distinct seasons. These forests consist of deciduous trees, which shed their leaves in the fall, and evergreen coniferous trees. They are known for their vibrant autumn colors and are important for timber production.

3.   Boreal   Forests: Boreal forests, also known as taiga, are found in the northern hemisphere, primarily in Canada, Russia, and Scandinavia. They are characterized by cold temperatures, long winters, and short growing seasons. Boreal forests are dominated by coniferous trees, such as spruce and pine, and provide critical habitat for many migratory bird species.

4. Mediterranean Forests: Mediterranean forests are found in regions with a Mediterranean climate, characterized by hot, dry summers and mild, wet winters. These forests consist of a mix of evergreen and deciduous trees, such as oak and olive. They are known for their rich biodiversity and are at risk due to increasing droughts and wildfires.

5. Montane Forests: Montane forests are found in mountainous regions, typically at higher elevations. They are characterized by cooler temperatures and higher rainfall compared to surrounding lowland areas. Montane forests are home to unique species adapted to the challenging mountainous environment and provide important water catchment areas.

Understanding the characteristics of different forest types is crucial for conservation biologists. By recognizing the specific needs of each ecosystem, they can develop effective strategies to protect and restore forest habitats. Conservation biology aims to maintain the ecological integrity of forests while allowing sustainable use of their resources.

Whether you are a nature enthusiast, a student, or someone interested in conservation biology, exploring the hidden world of forests is a fascinating journey. By understanding forest types and their characteristics, we can appreciate the incredible diversity of life they support and work towards preserving these vital ecosystems for future generations.

Biodiversity in Forest Ecosystems

Forests are not just a collection of trees; they are complex ecosystems teeming with life. Within the realm of forests, biodiversity flourishes, providing a rich tapestry of plants, animals, and microorganisms that work together to create a balanced and sustainable environment. In this subchapter, we will delve into the fascinating world of biodiversity in forest ecosystems and understand its significance in the field of conservation biology.

Biodiversity, simply put, refers to the variety of life forms found in a particular habitat or ecosystem. Forests, with their diverse range of habitats, serve as hotspots for biodiversity. From towering ancient trees to tiny insects, every organism plays a crucial role in maintaining the delicate balance of the forest ecosystem. By understanding and conserving this biodiversity, we can ensure the long-term survival of these complex ecosystems.

One key aspect of biodiversity in forest ecosystems is the incredible array of plant species. Forests are home to countless types of trees, shrubs, ferns, and other plants, each with its unique set of adaptations and ecological interactions. These plants not only provide habitat and food for countless animal species but also contribute to the overall health of the ecosystem by regulating climate, purifying air and water, and preventing soil erosion.

Animal biodiversity in forests is equally remarkable. From the majestic predators such as tigers and bears to the elusive birds and small mammals, forests provide a haven for a vast array of wildlife. These animals rely on the forest for food, shelter, and breeding grounds, forming intricate food webs and ecological relationships. The loss of even a single species can have far-reaching consequences, disrupting the delicate balance of the entire ecosystem.

Microorganisms, although often overlooked, also play a critical role in forest biodiversity. Fungi, bacteria, viruses, and other microorganisms are essential for nutrient cycling, decomposition, and maintaining soil health. They form symbiotic relationships with plants, helping them absorb essential nutrients and defend against pathogens. Without these tiny organisms, forest ecosystems would not be able to function effectively.

Conservation biology aims to protect and restore biodiversity, and forests are at the forefront of these efforts. By recognizing the value of biodiversity in forest ecosystems, we can work towards sustainable forest management practices, protect endangered species, and restore damaged habitats. Engaging everyone in these conservation efforts is crucial, as biodiversity loss impacts not only forests but also our own well-being.

In conclusion, biodiversity in forest ecosystems is a fascinating and intricate web of life. From the towering trees to the smallest microorganisms, every organism has a role to play in maintaining the health and balance of the forest ecosystem. By understanding and conserving this biodiversity, we can ensure the long-term survival of these invaluable ecosystems and contribute to the field of conservation biology. So, let us embark on a journey to explore and appreciate the hidden world of forest conservation biology together.

Interactions between Organisms in the Forest

In the enchanting realm of the forest, a complex web of interactions unfolds between its diverse inhabitants. From towering trees to tiny insects, every organism plays a crucial role in maintaining the delicate balance of this ecosystem. Understanding these interactions is key to unlocking the secrets of forest conservation biology.

One of the most fundamental interactions is between plants and animals. Forests provide a rich habitat for countless species, offering food, shelter, and protection. Birds and mammals, such as squirrels and deer, rely on the forest for sustenance and cover. In return, these animals assist in pollination and seed dispersal, ensuring the survival and diversity of plant species. For instance, hummingbirds, with their long beaks, are perfectly adapted to extract nectar from flowers, inadvertently transferring pollen from one bloom to another.

Beneath the forest floor lies another intricate interaction network: the soil microbiome. Fungi, bacteria, and other microorganisms work harmoniously to break down decaying organic matter, providing essential nutrients to plants. In return, plants exchange sugars with these microbial partners through their roots, forming a mutually beneficial relationship known as mycorrhizae. This symbiotic interaction enhances the forest's overall health and contributes to its resilience in the face of environmental challenges.

Predator-prey relationships are also prevalent in the forest, shaping the dynamics of its inhabitants. From wolves and deer to owls and mice, these interactions maintain population sizes and prevent any one species from dominating. Ecologists study these dynamics to understand how changes in predator or prey populations can ripple through the entire ecosystem. For example, a decline in top predators may result in a surge of herbivores, leading to overgrazing and detrimental effects on plant diversity.

Interactions in the forest extend beyond the realm of animals and plants. Forest fires, once considered destructive, have been revealed as an essential ecological process. Some tree species have even adapted to rely on wildfires to release their seeds or clear out competing vegetation. These fires also play a vital role in maintaining forest structure and rejuvenating the ecosystem, promoting biodiversity and preventing the dominance of a single species.

Understanding the intricate web of interactions between organisms in the forest is crucial for effective conservation biology. By conserving the forest and its inhabitants, we can protect countless species, maintain ecological balance, and preserve the hidden world that sustains us all. Whether you are an aspiring conservation biologist or simply a nature lover, delving into the complexities of these interactions will deepen your appreciation for the wonders of the forest and inspire you to contribute to its preservation.

The Role of Forests in Climate Regulation

Forests are often referred to as the lungs of our planet, and for good reason. These vast expanses of trees play a crucial role in regulating our climate and maintaining the delicate balance of our global ecosystem. In this subchapter, we will explore the significance of forests in climate regulation, highlighting their contribution to carbon sequestration, atmospheric stability, and climate mitigation.

To begin with, forests are one of the most effective natural carbon sinks on Earth. Through the process of photosynthesis, trees absorb carbon dioxide from the atmosphere and store it in their trunks, branches, and roots. This carbon sequestration helps to reduce the levels of greenhouse gases in the atmosphere, mitigating the impacts of climate change. In fact, forests are estimated to absorb approximately 2.6 billion metric tons of carbon dioxide each year, making them vital allies in our fight against global warming.

Furthermore, forests play a crucial role in maintaining atmospheric stability. The dense canopies of trees act as natural filters, intercepting rainfall and reducing the impact of heavy downpours. By doing so, forests help to regulate the water cycle, preventing floods and soil erosion. Additionally, the evapotranspiration process, whereby trees release water vapor into the atmosphere, helps to cool the surrounding air and regulate local temperatures. These mechanisms not only benefit the ecosystems within forests but also have far-reaching impacts on the climate patterns of entire regions.

In terms of climate mitigation, forests have the potential to play a significant role. Deforestation, unfortunately, contributes to approximately 15% of global greenhouse gas emissions. Therefore, protecting existing forests and restoring degraded ones is crucial in reducing these emissions. Forest conservation efforts, such as sustainable logging practices and reforestation initiatives, can help to not only preserve biodiversity but also mitigate climate change by increasing carbon sequestration.

In conclusion, forests are indispensable in climate regulation. Their ability to sequester carbon, maintain atmospheric stability, and mitigate climate change makes them essential components of our global ecosystem. As individuals interested in conservation biology, it is our collective responsibility to recognize the importance of forests and take action to preserve and restore them. Whether it is supporting sustainable forestry practices or participating in reforestation initiatives, every one of us has a role to play in safeguarding these invaluable ecosystems for future generations.

Chapter 3: The Science of Forest Conservation

The Basics of Conservation Biology

Conservation biology is a multidisciplinary field that aims to understand and protect Earth's biodiversity. It encompasses various scientific disciplines, including ecology, genetics, anthropology, and policy making. The field focuses on studying the relationships between organisms and their environment, with the ultimate goal of conserving and restoring ecosystems.

One of the fundamental principles of conservation biology is the understanding that all living organisms are interconnected and rely on each other for survival. This concept, known as biodiversity, refers to the variety of life forms found on Earth. Biodiversity is not only important for the survival of different species but also for the overall health and stability of ecosystems.

Conservation biologists strive to identify and address the causes of biodiversity loss. Habitat destruction, pollution, climate change, and overexploitation of natural resources are some of the major threats to biodiversity. By understanding these threats, conservation biologists can develop strategies to mitigate their impact on ecosystems.

One of the key tools in conservation biology is the concept of protected areas. These are designated regions that are managed to preserve their biodiversity and ecological processes. Protected areas can take various forms, such as national parks, wildlife refuges, or marine reserves. These areas provide a safe haven for many species and allow them to thrive without human interference.

Another important aspect of conservation biology is the study of endangered species. Endangered species are those that are at risk of extinction. Conservation biologists work to identify and protect these species by implementing conservation plans, breeding programs, and habitat restoration initiatives.

Conservation biology is not limited to the efforts of scientists and researchers alone. It requires the active participation of everyone, regardless of their background or expertise. Each individual can contribute to conservation efforts by making simple lifestyle changes, such as reducing waste, supporting sustainable practices, and advocating for environmental policies.

In summary, conservation biology is a vital field that aims to protect Earth's biodiversity and ensure the long-term survival of our planet. It encompasses various scientific disciplines and relies on the collective efforts of individuals from all walks of life. By understanding the basics of conservation biology and taking action, each and every one of us can contribute to the preservation of our natural world for future generations.

Ecological Principles in Forest Conservation

In a world grappling with climate change and habitat destruction, there is an urgent need to understand and implement effective strategies for forest conservation. Forests, the lungs of our planet, house an incredible array of species and provide crucial ecosystem services such as carbon sequestration, water regulation, and soil stabilization. To ensure the long-term survival of forests and the species within them, it is essential to apply ecological principles in forest conservation.

One key principle is biodiversity conservation. Forests are incredibly diverse ecosystems, harboring an astounding variety of plants, animals, and microorganisms. Preserving this biodiversity is crucial for maintaining the resilience of forests and their ability to withstand disturbances. By protecting the diversity of species within a forest, we can safeguard its ecological functions and ensure the survival of both common and rare species.

Another essential principle is the maintenance of ecological processes. Forests are complex systems where various ecological processes, such as nutrient cycling, pollination, and seed dispersal, interact to sustain life. Disruptions to these processes can have far-reaching consequences for the health and stability of the forest ecosystem. By understanding and preserving these processes, we can better manage and restore degraded forests, ensuring their long-term viability.

Furthermore, the concept of habitat connectivity plays a crucial role in forest conservation. Forests are often fragmented by human activities, leading to isolated patches that compromise species mobility and genetic exchange.

Maintaining or restoring connectivity between forest fragments is essential to facilitate the movement of species and enable the recolonization of disturbed areas. By creating corridors and protected areas that connect different forest patches, we can enhance biodiversity conservation and promote landscape-level ecological resilience.

Lastly, forest conservation should consider the involvement and empowerment of local communities. Indigenous peoples and local communities have valuable traditional knowledge and a deep understanding of forest ecosystems. Their involvement in conservation efforts ensures the sustainable management of forests, as their livelihoods and cultural practices are intimately linked to these ecosystems. By engaging and empowering local communities, we can foster a sense of ownership and shared responsibility for forest conservation.

Understanding and implementing these ecological principles in forest conservation is crucial for the survival of forests and the multitude of species that depend on them. By preserving biodiversity, maintaining ecological processes, promoting habitat connectivity, and involving local communities, we can work towards a sustainable future where forests thrive and continue to provide essential benefits for everyone. Only through collective action and a deep appreciation for the hidden world of forest conservation biology can we ensure the preservation of these invaluable ecosystems for generations to come.

Tools and Techniques for Forest Monitoring

In the realm of conservation biology, monitoring the health and well-being of forests is of paramount importance. Forests are not just a collection of trees; they constitute intricate ecosystems that support an array of species and provide vital services to our planet. To ensure their long-term survival, it is crucial to employ various tools and techniques for effective forest monitoring.

One primary tool used in forest monitoring is remote sensing. Satellites equipped with advanced sensors can capture high-resolution images of forests from space, enabling scientists to analyze changes in vegetation cover, deforestation rates, and forest fragmentation. This invaluable data provides insights into the extent and pace of forest loss, helping conservationists to identify areas of concern and prioritize efforts for protection and restoration.

Ground-based monitoring techniques complement remote sensing data, offering a more detailed perspective on forest health. One widely used method is biodiversity monitoring, which involves surveys of plant and animal species within a forest ecosystem. These surveys help scientists understand the diversity and abundance of species, identify threatened or endangered ones, and assess the overall ecological balance of the forest.

Another essential technique in forest monitoring is the use of acoustic monitoring devices. By recording sounds emitted by forest-dwelling animals, such as birds and mammals, researchers can gather valuable information about their population density, behavior, and habitat preferences. This data contributes to understanding the ecological health of the forest and can aid in identifying potential conservation hotspots.

In addition to these tools, citizen science initiatives play a significant role in forest monitoring. By engaging the general public in data collection, citizen science projects allow individuals from all walks of life to contribute to scientific research. Through smartphone applications, volunteers can report observations of tree species, invasive plants, and even the presence of wildlife in their local forests. This collective effort enhances the quality and quantity of data available for forest monitoring and strengthens public awareness and involvement in conservation efforts.

To effectively monitor forests, it is essential to combine these tools and techniques while considering the specific conservation goals and challenges of each region. By harnessing the power of remote sensing, ground-based surveys, acoustic monitoring, and citizen science, conservation biologists can gather comprehensive data on forest ecosystems. This knowledge forms the foundation for evidence-based decision-making and the implementation of effective conservation strategies.

The tools and techniques discussed here empower everyone - from scientists to local communities - to contribute to forest monitoring and conservation. By understanding the hidden world of forests and actively monitoring their health, we can work together to ensure the preservation of these vital ecosystems for the benefit of all species, including our own.

Data Collection and Analysis in Forest Conservation Biology

In the realm of forest conservation biology, the collection and analysis of data play a vital role in understanding and protecting the hidden world of forests. By examining the intricate web of interactions among various organisms and their environment, scientists can develop effective strategies to conserve these invaluable ecosystems. This subchapter aims to shed light on the important process of data collection and analysis in forest conservation biology, presenting it in a way that is accessible to everyone interested in the field of conservation biology.

Data collection in forest conservation biology involves gathering information on a wide range of factors, including species diversity, population dynamics, habitat quality, and ecosystem functioning. This information is obtained through various methods, such as field surveys, remote sensing techniques, and genetic analysis. Field surveys allow scientists to observe and record the presence and abundance of different species, as well as their behaviors and interactions. Remote sensing techniques, such as satellite imagery and aerial surveys, provide a broader perspective, enabling researchers to assess large-scale changes in forest cover and habitat fragmentation. Genetic analysis helps to uncover the genetic diversity and population structure of species, providing insights into their evolutionary history and potential for adaptation.

Once the data is collected, it is subjected to rigorous analysis to extract meaningful patterns and insights. Statistical techniques are employed to identify trends, relationships, and correlations among variables. This analysis helps scientists understand the impacts of human activities, such as deforestation and climate change, on forest ecosystems. It also aids in predicting future scenarios and evaluating the effectiveness of conservation efforts.

The data collected and analyzed in forest conservation biology not only contributes to scientific knowledge but also informs policy-making and management decisions. By understanding the complex dynamics of forest ecosystems, stakeholders can develop sustainable practices and policies that balance human needs with the conservation of biodiversity. Furthermore, this data serves as a baseline for long-term monitoring and assessment, enabling scientists to track changes in forest ecosystems over time and adapt conservation strategies accordingly.

In conclusion, data collection and analysis are essential components of forest conservation biology. By employing various methods and techniques, scientists are able to gather valuable information about forest ecosystems and understand their intricate workings. This knowledge allows for the development of effective conservation strategies, ensuring the preservation of these invaluable habitats for future generations.

Chapter 4: Forest Conservation Strategies

Protected Areas and Forest Reserves

In this subchapter, we delve into the fascinating world of protected areas and forest reserves, shedding light on their significance in the field of conservation biology. Whether you are an environmental enthusiast, a student, or simply someone curious about the hidden world of forests, this chapter aims to provide you with an accessible and engaging overview.

Protected areas are designated regions that are managed to preserve their ecological integrity and biodiversity. These areas play a crucial role in safeguarding our planet's natural resources and maintaining the delicate balance of ecosystems. Forest reserves, on the other hand, are specific protected areas that focus on the conservation of forest ecosystems. Together, they form a vital part of the global effort to combat habitat destruction, deforestation, and species extinction.

One key aspect of protected areas and forest reserves is their ability to act as safe havens for endangered species. By providing a sanctuary where diverse flora and fauna can thrive undisturbed, these areas help to prevent the loss of biodiversity. Moreover, protected areas serve as stepping stones for the movement and migration of species, enabling them to adapt to changing environmental conditions.

Another important benefit of protected areas is their contribution to climate change mitigation. Forest reserves, in particular, play a crucial role in carbon sequestration, helping to reduce greenhouse gas emissions. Forests act as carbon sinks, absorbing vast amounts of carbon dioxide and releasing oxygen through the process of photosynthesis. By protecting these areas, we can preserve their vital role in mitigating climate change and maintaining a stable climate system.

Moreover, protected areas and forest reserves offer numerous recreational and educational opportunities. They serve as natural classrooms, allowing people of all ages to learn about the intricate web of life and the importance of conservation. Visitors can engage in activities such as hiking, bird-watching, and nature photography, fostering a deeper connection with the natural world and inspiring a sense of responsibility towards its preservation.

In conclusion, protected areas and forest reserves are integral to the field of conservation biology. They provide a refuge for endangered species, contribute to climate change mitigation, and offer educational and recreational opportunities. By understanding and appreciating the value of these areas, we can actively participate in their conservation and ensure the preservation of our planet's natural heritage for future generations. So, let us embark on this journey into the hidden world of forests and discover the wonders that await us within these protected areas and forest reserves.

Sustainable Forest Management Practices

In this subchapter, we delve into the realm of sustainable forest management practices, exploring the vital role they play in the field of conservation biology. Forests are not only a mesmerizing part of our natural world but also serve as critical ecosystems that support a wide range of plant and animal species. Understanding the importance of sustainable practices is essential for everyone, as it empowers us to contribute towards the preservation of these invaluable habitats.

Sustainable forest management refers to the responsible use and conservation of forests to ensure their long-term health and productivity. It encompasses a variety of strategies aimed at balancing human needs with the protection and restoration of forest ecosystems. By implementing sustainable practices, we can mitigate deforestation, promote biodiversity conservation, and safeguard the many ecological services forests provide.

One key aspect of sustainable forest management is the concept of selective logging. Rather than clear-cutting entire forests, selective logging involves carefully choosing which trees to harvest, focusing on mature trees while leaving younger ones to continue growing. This approach minimizes the impact on the overall forest structure and biodiversity, allowing ecosystems to regenerate and thrive.

Additionally, sustainable forest management emphasizes the importance of reforestation and afforestation efforts. Reforestation involves replanting trees in areas where they have been cut down, while afforestation involves establishing forests in areas that were previously devoid of trees. These practices help combat deforestation and ensure the continuity of forest ecosystems, providing habitat for countless species and mitigating climate change by absorbing carbon dioxide.

Another crucial aspect of sustainable forest management is the inclusion of local communities and indigenous peoples in decision-making processes. Recognizing their traditional knowledge and rights, involving them in forest management fosters a sense of ownership and responsibility towards these ecosystems. This approach promotes sustainable practices while respecting the social and cultural values associated with forests.

By adopting sustainable forest management practices, we can create a harmonious balance between human needs and the conservation of forest ecosystems. It is crucial for everyone, from policymakers to individuals, to understand the importance of sustainable practices and actively contribute to their implementation. Together, we can ensure the preservation of these hidden worlds, protecting the biodiversity, ecosystem services, and natural beauty that forests offer for generations to come.

Reforestation and Afforestation Programs

In recent years, there has been a growing concern over the state of our planet's forests and the impact their destruction has on the environment. The loss of forests not only leads to a significant loss of biodiversity but also contributes to climate change, soil erosion, and the disruption of local communities. In response to these challenges, reforestation and afforestation programs have emerged as key solutions in the field of conservation biology.

Reforestation refers to the process of replanting trees in areas where forests have been harvested or destroyed. This practice aims to restore the ecological balance and functionality of the affected ecosystems. Reforestation programs typically involve careful planning and monitoring to ensure that the right tree species are chosen, taking into account the specific ecological conditions of the area. These programs also prioritize the use of native tree species to promote biodiversity and provide habitat for wildlife.

Afforestation, on the other hand, involves the establishment of new forests in areas that were previously devoid of trees. This practice is often employed in regions that have experienced severe deforestation or where the natural process of forest regeneration is hindered. Afforestation programs focus on selecting tree species that are well-suited to the local climate and soil conditions, with the goal of creating sustainable, self-reliant forests.

Both reforestation and afforestation programs offer numerous benefits for the environment and society as a whole. In addition to sequestering carbon dioxide and mitigating climate change, forests provide essential ecosystem services such as water filtration, soil stabilization, and the regulation of local temperatures. Moreover, these programs contribute to the preservation of biodiversity by restoring habitat for countless plant and animal species.

To ensure the success and long-term sustainability of reforestation and afforestation programs, it is crucial to involve local communities and stakeholders. Engaging and empowering these groups can help foster a sense of ownership and responsibility, leading to better preservation and management of the newly established forests. Additionally, these programs can provide opportunities for local communities to generate income through sustainable forest management practices, such as eco-tourism or the production of non-timber forest products.

In conclusion, reforestation and afforestation programs play a vital role in mitigating the negative impacts of deforestation and restoring the health of our planet's forests. By actively participating in these initiatives, we can contribute to the preservation of biodiversity, combat climate change, and ensure a sustainable future for ourselves and future generations.

Conservation through Community Engagement

In the realm of conservation biology, it has become increasingly clear that the key to successful and sustainable forest conservation lies in engaging local communities. The traditional top-down approach, where conservation efforts are imposed on communities without their active involvement, has proven to be ineffective in the long run. It is crucial to recognize that conservation is not just the responsibility of a few experts or organizations but something that should involve and benefit everyone.

Community engagement in forest conservation brings numerous advantages. Firstly, it fosters a sense of ownership and responsibility among the local people. When communities are actively engaged in the decision-making process and are given the opportunity to contribute their knowledge and experiences, they develop a deeper connection with the forest and a vested interest in its preservation. This sense of ownership leads to increased commitment and dedication in protecting the forest for future generations.

Furthermore, involving local communities in conservation efforts allows for the integration of traditional ecological knowledge. Indigenous communities, who have lived in harmony with the forest for centuries, possess a wealth of knowledge about the ecosystem and its inhabitants. By incorporating their wisdom and practices into conservation strategies, we can tap into centuries-old sustainable practices and enhance the effectiveness of our efforts.

Community engagement also promotes socio-economic development. By involving local communities in forest conservation projects, we can create alternative livelihood options that are both environmentally friendly and economically viable. This reduces the pressure on forests for resources such as timber and encourages sustainable practices like ecotourism, agroforestry, and non-timber forest product collection. Such initiatives not only provide economic benefits but also empower communities to take control of their own development.

To achieve successful community engagement, it is imperative to build strong partnerships between conservation organizations, researchers, and local communities. The collaboration should be based on mutual respect, trust, and shared goals. Capacity-building programs should be implemented to empower communities with the necessary skills and knowledge to actively participate in conservation efforts.

In conclusion, conservation through community engagement is not only essential but imperative for the long-term success of forest conservation. By involving local communities, we can ensure sustainable management practices, preserve traditional ecological knowledge, foster a sense of ownership, and promote socio-economic development. By recognizing that conservation is the responsibility of everyone, we can work towards a future where forests thrive and benefit not just the environment but also the communities that depend on them.

Chapter 5: Threats to Forest Conservation

Deforestation and its Causes

In today's rapidly changing world, one of the most pressing environmental issues we face is deforestation. The destruction of forests worldwide is a cause for great concern, as it not only impacts the biodiversity and the health of our planet but also threatens the livelihoods of millions of people who depend on forests for their survival.

Deforestation, simply put, refers to the permanent removal of trees from forests. While forests play a crucial role in maintaining the balance of our ecosystems, providing habitat for countless species, and acting as carbon sinks, they are being cleared at an alarming rate.

Several factors contribute to deforestation, and it is important for everyone to understand these causes to effectively address this global crisis. One major cause is commercial logging, driven by the demand for timber products. Unsustainable logging practices often result in the clear-cutting of vast areas of forest, leaving behind barren landscapes that take years, if not decades, to recover.

Another significant driver of deforestation is agriculture, particularly the expansion of large-scale commercial farming. As the global population continues to grow, the demand for food crops and livestock increases. This leads to the conversion of forests into agricultural land, resulting in the loss of invaluable biodiversity and the disruption of delicate ecosystems.

Furthermore, the extraction of natural resources, such as minerals and oil, also contributes to deforestation. These activities often involve the construction of roads and infrastructure, which opens up once inaccessible areas of forests to further exploitation.

Deforestation is also fueled by the need for land for human settlements and urbanization. The ever-expanding cities and towns require more space, leading to the clearing of forests for housing, industrial parks, and infrastructure development.

Lastly, climate change exacerbates deforestation as rising temperatures, and altered precipitation patterns increase the vulnerability of forests to pests, diseases, and wildfires. These disturbances weaken the resilience of forests, making them more susceptible to degradation.

To address the issue of deforestation, it is imperative for everyone, regardless of their background or profession, to understand the causes and implications of this destructive practice. By raising awareness about deforestation and its consequences, we can encourage individuals to make sustainable choices in their daily lives and support policies that prioritize forest conservation.

"The Hidden World: Exploring Forest Conservation Biology for Everyone" aims to provide a comprehensive understanding of deforestation and its causes. By delving into the intricate relationship between forests and conservation biology, this book empowers readers to take action and become advocates for forest preservation. Together, we can work towards a sustainable future where forests are protected, and the hidden world within them thrives.

Illegal Logging and Timber Trade

The issue of illegal logging and the timber trade has become a pressing concern in the field of conservation biology. This subchapter aims to shed light on the detrimental effects of these activities on our forests and the urgent need for action.

Illegal logging refers to the removal, processing, and trade of timber that violates national laws or international agreements. It is estimated that illegal logging accounts for a significant portion of the global timber trade, with devastating consequences for both the environment and local communities.

The impacts of illegal logging are far-reaching. Forests, which are vital ecosystems supporting countless plant and animal species, are being destroyed at an alarming rate. This loss of habitat threatens the survival of numerous species, some of which may be on the brink of extinction. Additionally, deforestation contributes to climate change by releasing vast amounts of carbon dioxide into the atmosphere.

In addition to its environmental consequences, illegal logging also has social and economic repercussions. Local communities that depend on forests for their livelihoods are often marginalized and left impoverished. The trade in illegal timber fuels corruption, undermines governance, and perpetuates organized crime networks.

Addressing the issue of illegal logging and timber trade requires a multi-faceted approach. It is crucial to strengthen law enforcement, improve governance, and promote sustainable forest management practices. International cooperation and partnerships between governments, NGOs, and local communities are essential to combat this global problem.

Furthermore, raising awareness among the general public is crucial. Each individual can play a role in combating illegal logging by making conscious choices when purchasing wood products. Supporting certified sustainable timber and demanding transparency in the supply chain can contribute to the reduction of illegal logging.

Conservation biologists have a vital role to play in this fight. By conducting research, advocating for stronger policies, and collaborating with stakeholders, they can contribute to the protection and preservation of our forests. The field of conservation biology offers a wealth of knowledge and tools that can be used to address the complex challenges posed by illegal logging and the timber trade.

In conclusion, illegal logging and the timber trade pose significant threats to our forests, biodiversity, and communities. By understanding the extent of the problem and taking action, we can work towards a more sustainable future. Conservation biologists, along with everyone else, have a responsibility to contribute to the conservation and protection of our precious forest ecosystems.

Forest Fragmentation and Habitat Loss

In the vast expanse of our planet's forests lies a hidden world of extraordinary biodiversity and intricate ecosystems. These forests are not only home to countless species of plants and animals but also play a crucial role in maintaining the balance of our planet's climate. However, the hidden world is under threat, and one of the most significant challenges it faces is forest fragmentation and habitat loss.

Forest fragmentation refers to the process of breaking up large, continuous forest areas into smaller, isolated patches. This fragmentation occurs primarily due to human activities such as agriculture, urbanization, and infrastructure development. As a result, the once-uninterrupted habitats become fragmented islands surrounded by a sea of human-modified landscapes. This loss of connectivity poses serious threats to the survival of many species.

Habitat loss is closely linked to forest fragmentation, as the destruction of forests directly results in the loss of habitat for countless species. As forests are cleared for agriculture or logging, species that depend on these habitats for food, shelter, and breeding are left with limited options and face increased competition for resources. This can lead to declines in population sizes, reduced genetic diversity, and, in some cases, even local extinctions.

The consequences of forest fragmentation and habitat loss extend beyond the immediate impact on individual species. These processes disrupt the intricate web of ecological interactions within forests, affecting the overall ecosystem functioning. For example, the loss of top predators can result in an increase in prey species, leading to imbalances in the food chain. Additionally, fragmentation can increase the edge effects, exposing the forest interior to disturbances such as invasive species, pollution, and human encroachment.

Conservation biology plays a vital role in addressing the challenges posed by forest fragmentation and habitat loss. Through scientific research, conservationists aim to understand the impacts of these processes on forest ecosystems and develop strategies to mitigate their effects. This includes creating corridors to reconnect fragmented habitats, implementing sustainable land-use practices, and advocating for protected areas.

However, the responsibility of conserving forests does not solely lie with scientists and conservationists. Each and every one of us has a role to play in protecting these invaluable ecosystems. By making conscious choices in our daily lives, such as supporting sustainable products and reducing our ecological footprint, we can contribute to the preservation of forests and the countless lives they sustain.

The hidden world of forests is a treasure worth protecting. By understanding and addressing the challenges of forest fragmentation and habitat loss, we can ensure that future generations will have the opportunity to explore and appreciate the wonders of these diverse and vital ecosystems. Let us come together and embrace our collective responsibility to conserve our forests for everyone.

Climate Change and its Impact on Forests

Forests play a vital role in maintaining the delicate balance of our planet's ecosystem. They are not only home to countless species of plants and animals but also provide us with numerous essential resources. However, in recent years, forests have been facing an unprecedented threat due to climate change.

Climate change refers to long-term shifts in temperature, precipitation patterns, and other climatic conditions. These changes have been primarily driven by human activities, such as the burning of fossil fuels and deforestation. As a result, our forests are experiencing significant impacts, which have far-reaching consequences for both the environment and human society.

One of the most noticeable impacts of climate change on forests is the alteration of their geographical distribution. As temperatures rise, certain tree species are no longer able to thrive in their traditional habitats. This leads to a phenomenon known as "forest migration," where trees move towards higher latitudes or higher elevations in search of suitable conditions. Unfortunately, this migration is often hindered by human settlements, infrastructure, and other barriers, causing fragmentation and loss of forest cover.

Another key effect of climate change is the increased frequency and intensity of natural disasters. Forests act as natural buffers against extreme weather events such as hurricanes, floods, and droughts. However, with climate change, these events become more frequent and severe, leading to widespread forest destruction. The loss of trees not only reduces carbon sequestration capacity but also leaves the soil vulnerable to erosion, further exacerbating the impact on ecosystems.

Moreover, climate change also affects the delicate balance between plants and their pollinators, disrupting the reproductive cycle of many tree species. With warmer temperatures, the timing of flowering and the emergence of pollinators may become unsynchronized, jeopardizing the reproduction and survival of certain tree species. This, in turn, can lead to a decline in forest biodiversity and disrupt the intricate web of ecological interactions within the forest ecosystem.

The consequences of climate change on forests are not limited to the environment. Forests provide numerous ecosystem services that are essential for human well-being, such as water filtration, carbon storage, and recreation. The degradation and loss of forests can have severe socio-economic implications, including the loss of livelihoods for local communities and an increase in the vulnerability of communities to natural disasters.

Addressing the impacts of climate change on forests requires urgent global action. Efforts must be made to reduce greenhouse gas emissions, promote sustainable land use practices, and protect and restore forest ecosystems. Moreover, it is crucial to raise awareness among individuals and communities about the importance of forests and their role in mitigating climate change.

In conclusion, climate change poses a significant threat to the world's forests and the countless species that depend on them. Understanding the impacts of climate change on forests is crucial for everyone, especially those interested in conservation biology. By working together, we can take meaningful steps towards mitigating climate change and ensuring the long-term survival of our precious forest ecosystems.

Chapter 6: Case Studies in Forest Conservation

Successful Forest Conservation Projects

In the realm of conservation biology, numerous successful forest conservation projects have been carried out worldwide, showcasing the power of collective efforts to protect and preserve our invaluable forest ecosystems. These projects, undertaken by organizations, communities, and individuals, have made significant contributions towards safeguarding the intricate web of life within forests and supporting sustainable development.

One such notable project is the Amazon Rainforest Conservation Initiative. The Amazon rainforest, often referred to as the "lungs of the Earth," faces numerous threats, including deforestation and illegal logging. The initiative involves a collaborative effort between local communities, NGOs, and governments to combat these threats. Through extensive reforestation programs, implementation of sustainable land-use practices, and the establishment of protected areas, this project has successfully slowed down deforestation rates in the Amazon, preserving biodiversity and mitigating climate change.

Another remarkable endeavor is the Great Green Wall project in Africa. This ambitious project aims to combat desertification by creating a green belt of trees across the Sahel region. By planting a mix of indigenous tree species, local communities are restoring degraded land, preventing soil erosion, and providing habitats for diverse wildlife. Moreover, this project offers economic opportunities for communities through sustainable land management practices, such as agroforestry, which combines tree planting with agricultural practices.

Closer to home, the Redwood National and State Parks in California have been a triumph in forest conservation. These parks protect one of the world's tallest tree species, the iconic coast redwoods. Through active management strategies, including controlled burns and the removal of invasive species, these forests have been restored to their natural state. Additionally, educational programs and visitor centers have been established to increase awareness and appreciation for these majestic forests, ensuring their long-term conservation.

These successful forest conservation projects demonstrate the importance of partnerships, engagement with local communities, and the integration of scientific knowledge with traditional practices. By recognizing the intrinsic value of forests and their critical role in supporting biodiversity, regulating climate, and providing ecosystem services, these projects have paved the way for a more sustainable future.

In conclusion, the success stories in forest conservation inspire hope and provide valuable lessons for individuals, communities, and policymakers. Through collective action, we can contribute to the preservation of forests, not only for their intrinsic value but also for the countless benefits they provide to humanity. By learning from these projects, we can empower ourselves to become active participants in the conservation of our planet's most precious resource — the hidden world of forests.

Challenges and Lessons Learned from Forest Conservation Initiatives

Forest conservation initiatives play a crucial role in preserving the Earth's biodiversity, mitigating climate change, and ensuring sustainable resource management. However, these efforts are not without their fair share of challenges. In this subchapter, we will explore some of the key obstacles faced by forest conservation initiatives and the valuable lessons learned along the way.

One of the biggest challenges in forest conservation is the encroachment of human activities. Rapid urbanization, industrialization, and agricultural expansion often result in deforestation and habitat fragmentation. Balancing economic development with environmental conservation becomes a delicate task. However, through various projects and initiatives, valuable lessons have been learned. It is now widely acknowledged that involving local communities in decision-making processes, providing alternative livelihood options, and raising awareness about the importance of forests are essential for the success of any conservation endeavor.

Another significant challenge is the threat of illegal logging and poaching. The demand for timber, wildlife products, and medicinal plants drives illegal activities that undermine the conservation efforts. Forest conservation initiatives have recognized the need for stringent law enforcement, international cooperation, and community engagement to combat these illegal activities effectively. By partnering with local law enforcement agencies, empowering local communities to monitor and report illegal activities, and raising awareness about the impacts of illegal resource extraction, significant progress has been made in curbing these practices.

Climate change poses another pressing challenge to forest conservation. Rising temperatures, changing rainfall patterns, and increased frequency of extreme weather events affect the health and resilience of forests. Conservation initiatives have learned valuable lessons in implementing adaptive management strategies, such as assisted migration, reforestation with climate-resilient species, and the promotion of agroforestry practices. These approaches help forests adapt to changing environmental conditions and maintain their vital ecological functions.

Furthermore, forest conservation initiatives have realized the importance of collaboration and interdisciplinary approaches. Addressing complex challenges necessitates the involvement of various stakeholders, including scientists, policymakers, local communities, and NGOs. By fostering partnerships, sharing knowledge and resources, and integrating multiple perspectives, conservation efforts have become more holistic and effective.

In conclusion, forest conservation initiatives face numerous challenges, but valuable lessons have been learned along the way. By involving local communities, combating illegal activities, adapting to climate change, and fostering collaboration, these initiatives contribute to the preservation of forests and the protection of biodiversity. The lessons learned from these experiences can inspire and guide future conservation biology efforts, ensuring a sustainable and vibrant hidden world for everyone.

Forest Conservation Efforts at the Global Scale

Forests are invaluable ecosystems that provide numerous benefits to the planet and its inhabitants. They are home to millions of species, act as carbon sinks, regulate the climate, and provide essential resources for human societies. However, due to various human activities, forests are facing unprecedented threats, such as deforestation, habitat fragmentation, and climate change. In response to these challenges, global efforts to conserve forests have emerged, aiming to protect these vital ecosystems for future generations.

One of the primary global initiatives in forest conservation is the establishment of protected areas. These designated areas, often national parks or reserves, are crucial in safeguarding the biodiversity and ecological integrity of forests. Protected areas not only provide a safe haven for endangered species but also serve as important research sites to understand the complex dynamics of forest ecosystems. These areas also play a significant role in promoting ecotourism, generating income for local communities while simultaneously creating incentives for forest preservation.

Another essential aspect of global forest conservation efforts is sustainable forest management. This approach recognizes the importance of balancing human needs with ecological integrity. Sustainable management practices aim to extract forest resources in a way that minimizes negative impacts on the ecosystem, such as selective logging techniques and reforestation efforts. By adopting sustainable practices, we can ensure the long-term viability of forests and preserve their invaluable services.

International collaborations and agreements are also critical in promoting forest conservation at a global scale. The United Nations has played a pivotal role in this regard, particularly through its Sustainable Development Goals. Goal 15 specifically targets the protection, restoration, and sustainable use of terrestrial ecosystems, including forests. Additionally, the United Nations Framework Convention on Climate Change (UNFCCC) has recognized the importance of forests in mitigating climate change, leading to initiatives such as REDD+ (Reducing Emissions from Deforestation and Forest Degradation).

Moreover, public awareness and engagement are vital components of forest conservation efforts. Individuals from all walks of life can contribute to the cause by making conscious choices, such as supporting sustainable products, reducing their carbon footprint, and participating in reforestation programs. Education and outreach programs can also play a significant role in fostering a sense of environmental responsibility and inspiring a new generation of conservation biologists.

In conclusion, forest conservation efforts at the global scale are essential to preserve the invaluable services and biodiversity provided by these ecosystems. Through the establishment of protected areas, sustainable management practices, international collaborations, and public engagement, we can ensure the long-term survival of forests. As individuals, we all have a role to play in this endeavor, and by working together, we can create a world where forests thrive and continue to benefit everyone.

Local and Indigenous Perspectives on Forest Conservation

In the pursuit of forest conservation, it is essential to acknowledge and appreciate the invaluable knowledge and perspectives held by local and indigenous communities. These communities have long-standing relationships with the forests, rooted in traditional ecological knowledge and sustainable practices that have ensured the preservation of these precious ecosystems for generations.

Local and indigenous communities have a deep understanding of the forests' intricacies, biodiversity, and the delicate balance that must be maintained for their survival. This knowledge is often passed down through oral traditions and shared among community members, ensuring the continuity of their conservation efforts.

One of the key aspects of local and indigenous perspectives on forest conservation is the recognition of the interconnectedness between humans and the natural world. These communities perceive the forest not merely as a resource to exploit but as an entity that sustains their livelihoods, cultures, and spirituality. Their holistic approach to conservation encompasses social, economic, and environmental dimensions, ensuring the well-being of both humans and the forest ecosystem.

Furthermore, local and indigenous communities play a vital role in monitoring and managing forests. Their intimate knowledge enables them to identify changes in the forest dynamics, such as shifts in biodiversity, climate patterns, and the presence of invasive species. This knowledge, combined with modern scientific techniques, creates a powerful synergy that can enhance forest conservation efforts.

Engaging local and indigenous communities in forest conservation is not only ethically imperative but also crucial for the success of these initiatives. By involving communities in decision-making processes, their perspectives and interests can be integrated into conservation strategies, fostering a sense of ownership and stewardship. Moreover, local communities often possess the traditional practices and techniques that can contribute to sustainable forest management, including agroforestry, rotational farming, and selective logging.

In conclusion, local and indigenous perspectives on forest conservation bring a wealth of knowledge and wisdom that can enhance our understanding and approach to preserving these vital ecosystems. By recognizing the interconnectedness between humans and the natural world, and actively involving local communities in decision-making processes, we can foster a more inclusive and effective conservation approach. It is imperative that we embrace these perspectives and work collaboratively to ensure the long-term survival of our forests for the benefit of all — humans, wildlife, and the planet as a whole.

Chapter 7: The Role of Individuals in Forest Conservation

Everyday Actions for Forest Conservation

In our rapidly changing world, forests hold a vital role in preserving biodiversity, combating climate change, and providing countless benefits to both humans and wildlife. As individuals, we often underestimate the impact our everyday actions can have on forest conservation. However, by making conscious choices and adopting simple practices, we can all contribute to the preservation of these precious ecosystems.

1. Reduce, Reuse, Recycle: One of the fundamental steps in forest conservation is minimizing waste. By reducing our consumption, reusing products whenever possible, and recycling materials, we can reduce the demand for resources extracted from forests. This, in turn, helps reduce deforestation and habitat loss.

2. Choose Sustainable Products: When purchasing wood or paper-based products, opt for those certified by sustainable forestry initiatives like the Forest Stewardship Council (FSC). These certifications ensure that the products are sourced from responsibly managed forests, promoting the conservation of biodiversity and the rights of forest communities.

3. Plant Trees: Participate in tree-planting initiatives in your community or even in your own backyard. Trees not only provide oxygen, but they also serve as crucial habitats for numerous species. By planting native tree species, you can help restore and expand forest ecosystems.

4. Conserve Water: Conserving water is directly linked to forest conservation. Forests rely on rainfall and act as natural water filters, regulating water flow and maintaining water quality. By reducing water usage at home and preventing pollution, we can help protect forest ecosystems and the species that depend on them.

5. Support Forest Conservation Organizations: Donate your time, money, or resources to organizations dedicated to forest conservation. These organizations work tirelessly to protect and restore forests, conduct research, and empower local communities. Your support can make a significant difference in their efforts.

6. Educate Yourself and Others: Learn more about the importance of forests and the challenges they face. Share your knowledge with others to raise awareness and inspire action. By educating ourselves and others, we can create a collective consciousness that values and prioritizes forest conservation.

Remember, forest conservation is a shared responsibility. Regardless of our background or expertise, each of us has a role to play in protecting these invaluable ecosystems. By implementing these everyday actions, we can contribute to the preservation of forests for future generations and ensure a sustainable future for all. Together, we can make a difference in the world of conservation biology.

Getting Involved in Forest Conservation Organizations

In the quest to protect and preserve our precious forests, getting involved in forest conservation organizations can play a pivotal role. These organizations not only strive to raise awareness about the importance of forest conservation but also actively work towards implementing sustainable practices to ensure the longevity of our forests. If you have a passion for conservation biology and want to make a positive impact, here are some ways you can get involved in forest conservation organizations.

1. Volunteer Opportunities: Many forest conservation organizations offer volunteer programs that allow individuals to contribute their time and skills towards conservation efforts. Whether it's participating in tree planting events, conducting wildlife surveys, or assisting in habitat restoration projects, volunteering is a hands-on way to make a difference. These experiences can also provide valuable learning opportunities and help you develop a deeper understanding of forest ecosystems.

2. Join Membership Programs: Becoming a member of forest conservation organizations not only supports their work financially but also grants you access to a network of like-minded individuals. Membership programs often offer benefits such as newsletters, educational resources, and invitations to special events. By joining these programs, you can stay updated on the latest conservation initiatives and connect with experts in the field.

3. Fundraising and Donations: Forest conservation organizations heavily rely on donations to fund their research, conservation projects, and educational programs. If you have

the means, consider making a financial contribution to support their work. Additionally, you can organize fundraising events or campaigns to raise awareness and gather resources for specific conservation projects.

4. Advocacy and Education: Take an active role in spreading awareness about forest conservation by becoming an advocate for these organizations. Attend public meetings, participate in community outreach programs, and use social media platforms to educate others about the importance of forest conservation. By sharing your knowledge and passion, you can inspire others to take action and make a difference.

5. Career Opportunities: If you are deeply committed to the field of conservation biology, consider pursuing a career in forest conservation organizations. Many organizations offer employment opportunities in research, policy-making, and project management. By working directly in the field, you can contribute your expertise towards developing sustainable conservation strategies and shaping future policies.

Remember, every individual has the power to make a difference in forest conservation. By getting involved in forest conservation organizations, you can actively contribute towards protecting and preserving these vital ecosystems for future generations. Together, we can create a world where forests thrive, biodiversity flourishes, and the hidden wonders of the forest can be enjoyed by everyone.

Advocacy and Policy Change for Forest Protection

Forests are the lifeblood of our planet, providing numerous ecological, economic, and social benefits. Unfortunately, they are under constant threat from deforestation, unsustainable logging practices, and land conversion. As concerned individuals, it is crucial that we understand the importance of advocacy and policy change in protecting and conserving these invaluable ecosystems.

Advocacy serves as a powerful tool for raising awareness about the urgent need for forest protection. It involves educating and mobilizing individuals, communities, and governments to take action. By becoming advocates for forest conservation, we can help shape public opinion, influence decision-makers, and encourage sustainable practices.

One of the first steps in advocacy is understanding the complex web of policies that govern forest management. Policies play a pivotal role in determining the fate of our forests, and it is essential to be well-informed about existing regulations and their implications. By familiarizing ourselves with the policies in place, we can identify gaps and areas where change is necessary.

Policy change is a long and challenging process, but with collective effort, it is possible to bring about positive transformations. Advocacy campaigns can range from grassroots initiatives to international movements, depending on the issue at hand. These campaigns can include activities such as writing letters to policymakers, organizing protests, engaging in public consultations, or even using social media platforms to spread awareness. The goal is to create a groundswell of support that compels policymakers to take action.

Furthermore, engaging with local communities and indigenous groups is vital in advocating for forest protection. These communities often have intimate knowledge of the forests and are directly affected by any policy decisions. By involving them in the advocacy process, we can ensure that their voices are heard and their perspectives considered. Collaborating with these groups can lead to more inclusive and effective policy changes, as it combines scientific expertise with traditional ecological knowledge.

Ultimately, advocacy and policy change are essential components of conservation biology. They empower individuals and communities to take a stand against the destruction of our forests and promote sustainable practices. By working together, we can create a future where forests thrive, biodiversity is protected, and the benefits they provide are enjoyed by everyone.

"The Hidden World: Exploring Forest Conservation Biology for Everyone" aims to enlighten and inspire readers from all walks of life, emphasizing the significance of advocacy and policy change in forest protection. Whether you are a student, environmentalist, policymaker, or simply a concerned citizen, this subchapter underscores the role each one of us can play in safeguarding our forests for generations to come.

The Power of Education and Awareness in Forest Conservation

In the pursuit of sustaining our planet's valuable ecosystems, education and awareness play a pivotal role in the field of forest conservation biology. With forests facing unprecedented threats such as deforestation, habitat fragmentation, and climate change, it has become imperative for everyone to understand the importance of conserving these magnificent habitats and the diverse life they support.

Education is the key to promoting a deeper understanding of the complex web of life within forests. By providing individuals with the knowledge and tools to comprehend the intricate relationships between species, ecosystems, and human activities, we can empower them to make informed decisions and take action to protect our forests.

From an early age, education should instill a sense of wonder and appreciation for forests, nurturing a love for nature that will last a lifetime. By integrating forest conservation topics into school curricula, we can ensure that every student gains a fundamental understanding of the value of forests and their role in maintaining ecological balance. Educating children about the various threats faced by forests and the potential consequences of their destruction can inspire them to become future advocates for conservation.

However, education alone is not enough. Awareness is crucial in bridging the gap between knowledge and action. By raising awareness about forest conservation issues through various media platforms, public campaigns, and community engagement programs, we can reach a broader audience and inspire collective action. Through documentaries, articles, and social media, we can showcase the beauty and importance of forests, while highlighting the urgent need for their preservation.

Furthermore, fostering partnerships between conservation biologists, local communities, and policymakers is essential to ensure the implementation of effective conservation strategies. By involving local communities in forest conservation initiatives, we can benefit from their invaluable traditional knowledge and create a sense of ownership and responsibility towards these natural resources.

In summary, education and awareness are indispensable tools in the realm of forest conservation biology. By equipping individuals with knowledge, nurturing a sense of wonder, and fostering awareness, we can inspire a collective effort to protect and restore our forests. Together, we can create a future where forests thrive, ecosystems flourish, and the hidden wonders of the natural world are cherished and preserved for generations to come.

Chapter 8: The Future of Forest Conservation

Emerging Technologies in Forest Conservation

In the ever-evolving field of conservation biology, new technologies are constantly emerging to aid in the preservation and management of our planet's forests. These cutting-edge innovations have the potential to revolutionize the way we approach forest conservation, providing us with powerful tools to better understand, protect, and restore these invaluable ecosystems.

One of the most promising technologies in forest conservation is remote sensing. Advances in satellite imagery and aerial drones have allowed researchers to gather vast amounts of data about forests from a bird's eye view. This data can be used to monitor deforestation, track habitat loss, and assess the overall health of forest ecosystems. By analyzing this information, scientists can identify areas at high risk and target their conservation efforts more effectively.

Another emerging technology that holds great promise is DNA barcoding. This technique allows researchers to identify species by analyzing their unique genetic markers. In the context of forest conservation, DNA barcoding can help identify and protect endangered plant and animal species, as well as detect illegal logging and trade in protected tree species. By harnessing the power of DNA, conservationists can gain a comprehensive understanding of the biodiversity within a forest and design conservation strategies accordingly.

Furthermore, the advent of big data and artificial intelligence (AI) has opened up new possibilities in forest conservation. By integrating data from various sources, such as satellite imagery, climate records, and species distribution models, AI algorithms can provide real-time insights into the status and dynamics of forest ecosystems. This can help inform policy decisions, guide land-use planning, and predict the impacts of climate change on forests. Additionally, AI-powered technologies can assist in the efficient monitoring and management of protected areas, ensuring that they are adequately protected from human encroachment and disturbances.

Lastly, emerging technologies in forest restoration, such as tree-planting drones and precision forestry, offer innovative approaches to combat deforestation and restore degraded landscapes. These technologies can automate the process of reforestation, ensuring that trees are planted in the right locations and with optimal spacing, leading to more successful restoration efforts.

As we delve into the hidden world of forest conservation biology, it is essential for everyone to be aware of these emerging technologies and their potential to shape the future of forest conservation. By staying informed and embracing these advancements, we can all play a part in safeguarding the world's forests for future generations. Whether you are a scientist, policymaker, or simply someone who cares about the environment, understanding and supporting these technologies is crucial in our collective efforts to preserve the beauty and functionality of our forests.

Integrating Forest Conservation with Sustainable Development Goals

In recent years, the importance of sustainable development and environmental conservation has gained significant attention worldwide. As we strive to address the challenges posed by climate change, deforestation, and biodiversity loss, we must consider the crucial role that forests play in achieving sustainable development goals. This subchapter explores how integrating forest conservation with sustainable development goals can pave the way for a harmonious coexistence between human activities and nature.

Forests, often referred to as the lungs of our planet, offer numerous benefits to both humans and wildlife. They provide habitat for countless species, protect watersheds, and act as carbon sinks, mitigating the impacts of climate change. However, the indiscriminate exploitation of forests for timber, agriculture, and urbanization has resulted in alarming rates of deforestation, leading to the loss of biodiversity and disruption of ecosystem services. To reverse this trend, it is imperative to integrate forest conservation into our broader sustainable development agenda.

The concept of sustainable development aims to meet the needs of the present generation without compromising the ability of future generations to meet their own needs. By integrating forest conservation with sustainable development goals, we can strike a balance between economic growth, social well-being, and environmental protection. This integration requires a multidisciplinary approach that involves stakeholders from various sectors, including government agencies, non-profit organizations, and local communities.

One way to achieve this integration is through the implementation of sustainable forest management practices. This approach ensures that forests are managed in a way that meets current needs while preserving their capacity to provide resources and environmental services in the long term. Sustainable forest management involves practices such as selective logging, reforestation, and the protection of sensitive areas. By adopting these practices, we can maintain the ecological integrity of forests while also providing economic opportunities for local communities.

Furthermore, integrating forest conservation with sustainable development goals requires addressing the underlying drivers of deforestation, such as poverty and unsustainable agricultural practices. Enhancing livelihood opportunities for local communities, promoting sustainable agriculture, and providing alternative income sources can help alleviate the pressures on forests and reduce reliance on destructive activities.

In conclusion, integrating forest conservation with sustainable development goals is essential for achieving a sustainable and prosperous future for all. By recognizing the value of forests and implementing sustainable forest management practices, we can ensure the preservation of biodiversity, the protection of ecosystem services, and the well-being of both present and future generations. It is a collective responsibility to promote this integration and work towards a harmonious coexistence between human activities and the natural world. Let us strive to create a hidden world of forest conservation biology that is accessible and beneficial to everyone.

Addressing Global Forest Conservation Challenges

Forests are one of the most valuable ecosystems on our planet, providing numerous benefits for both humans and wildlife. However, in recent years, global forest conservation has faced numerous challenges that threaten the very existence of these vital ecosystems. In this subchapter, we will delve into the pressing issues surrounding forest conservation and explore potential solutions that can be implemented by everyone.

Deforestation, driven by human activities such as logging, agriculture expansion, and infrastructure development, remains a primary concern. Every year, millions of hectares of forests are lost, leading to the destruction of habitats, loss of biodiversity, and contributing to climate change. To address this challenge, it is crucial for everyone to understand the importance of forests and actively participate in their conservation.

One effective solution is promoting sustainable forestry practices. By adopting responsible logging techniques, we can ensure that the economic benefits of forests are balanced with long-term ecological sustainability. This can be achieved through the implementation of certification programs, such as the Forest Stewardship Council (FSC), which ensures that timber and forest products come from responsibly managed forests.

Another key challenge is the illegal wildlife trade, which often involves the exploitation of forest habitats. This trade not only threatens iconic species like elephants, tigers, and rhinos but also disrupts entire ecosystems. Addressing this issue requires robust law enforcement, community involvement, and promoting sustainable livelihoods for local communities to discourage their involvement in illegal activities.

Furthermore, addressing global forest conservation challenges requires international collaboration and policy frameworks. Governments, non-governmental organizations, and individuals must work together to establish protected areas, promote reforestation initiatives, and develop sustainable land-use policies. It is crucial for everyone to advocate for stronger environmental regulations and support organizations working on the ground to protect forests.

Education and awareness also play a vital role in forest conservation. By sharing knowledge about the ecological importance of forests and the threats they face, we can inspire more people to take action. Engaging with schools, organizing community events, and utilizing social media platforms can help spread the message and encourage individuals to make sustainable choices in their daily lives.

In conclusion, addressing global forest conservation challenges requires a collective effort from everyone. By promoting sustainable forestry practices, combating illegal wildlife trade, advocating for policy changes, and raising awareness, we can contribute to the protection and preservation of our precious forests. Each individual has a role to play, and by working together, we can ensure that future generations can continue to benefit from the hidden world of forests.

Hope for the Future: Inspiring Stories of Successful Forest Conservation

Forests are not just a collection of trees; they are intricate ecosystems that support a diverse range of plants, animals, and communities. However, over the years, rampant deforestation has posed a significant threat to these vital habitats and the countless species that call them home. But amidst the challenges, there is hope for the future of forest conservation. In this subchapter, we explore inspiring stories of successful forest conservation initiatives that have made a real difference in preserving these invaluable ecosystems.

One such story takes us to the heart of the Amazon rainforest, where a group of local communities came together to protect their ancestral lands from illegal logging and land encroachment. Through the formation of a grassroots organization, they implemented sustainable forestry practices, establishing community-managed reserves that safeguarded their forests and provided them with a sustainable source of income. The success of this initiative not only protected the biodiversity-rich Amazon but also empowered the indigenous communities, demonstrating the power of collective action.

Another remarkable tale comes from the Boreal Forest in Canada, where conservationists worked tirelessly to protect this vast wilderness from industrial exploitation. Through strategic partnerships with logging companies, they developed sustainable logging practices that minimized the environmental impact while allowing for economic development. This innovative approach to forest management not only preserved the integrity of the Boreal Forest but also provided a blueprint for sustainable resource extraction worldwide.

The subchapter also delves into the inspiring stories of reforestation efforts in devastated areas. One such example takes us to the Loess Plateau in China, where decades of intensive farming and erosion had turned the land into a barren wasteland. Through a massive reforestation project, the local communities transformed the landscape, planting millions of trees that halted soil erosion, restored water sources, and revived the region's biodiversity. This incredible feat showcases the resilience of nature and the power of human determination to restore what was once lost.

These stories of successful forest conservation serve as a beacon of hope for the future. They highlight the crucial role that individuals, communities, and organizations play in protecting and restoring our forests. By learning from these examples and implementing sustainable practices, we can ensure the preservation of our forests for future generations.

Whether you are a conservation biologist, an environmental enthusiast, or simply someone who cares about the well-being of our planet, these inspiring stories will captivate and motivate you. They remind us that it is not too late to make a difference and that by working together, we can create a better, greener future for all.

Chapter 9: Conclusion

Recap of Key Points

Throughout "The Hidden World: Exploring Forest Conservation Biology for Everyone," we have delved into the fascinating realm of conservation biology, shedding light on the intricate workings of forests and the urgent need to protect them. In this final subchapter, we will recapitulate the key points discussed, ensuring that these crucial lessons remain ingrained in the minds of every reader, regardless of their background or expertise.

First and foremost, we have emphasized the vital role that forests play in maintaining the delicate balance of our planet's ecosystems. Forests are not merely a collection of trees; they are complex habitats teeming with diverse flora and fauna, each species interdependent on the other. By understanding the intricate web of relationships within forests, we can grasp the urgency of conserving these invaluable ecosystems.

We have also explored the numerous threats faced by forests today. From deforestation and habitat fragmentation to climate change and invasive species, the challenges are immense. However, we have highlighted the power of collective action, emphasizing that each individual can make a difference. Whether through supporting sustainable consumer choices, participating in reforestation efforts, or advocating for stronger environmental policies, every action counts.

Furthermore, we have examined the innovative techniques employed by conservation biologists to study and protect forests. From utilizing remote sensing technologies to monitor forest cover changes to conducting genetic analyses to understand the health and diversity of tree populations, these tools have revolutionized our understanding of forest ecosystems. By embracing these advances, we can forge a path towards effective conservation strategies.

Additionally, we have stressed the importance of involving local communities in forest conservation efforts. Recognizing the intrinsic connections between forests and human livelihoods, it is crucial to engage communities as stakeholders in conservation initiatives. By empowering local people and incorporating traditional knowledge, we can foster sustainable practices that benefit both forests and human well-being.

Lastly, we have emphasized the need for ongoing research and education in the field of forest conservation biology. Our understanding of forests is constantly evolving, and it is imperative to remain updated with the latest scientific advancements. By promoting an inclusive and accessible approach to learning, we can inspire a new generation of conservation biologists who will carry the torch forward.

In conclusion, "The Hidden World: Exploring Forest Conservation Biology for Everyone" has provided a comprehensive overview of the critical issues surrounding forest conservation. By recapitulating the key points discussed, we hope to reinforce the urgency of protecting forests and inspire individuals from all walks of life to take action. Together, we can ensure the preservation of these magnificent ecosystems for generations to come.

The Importance of Continued Efforts in Forest Conservation

Forests are essential for the survival of all life forms on our planet. They provide us with clean air, fresh water, and are home to an incredible diversity of plants and animals. However, in recent years, the world's forests have been facing unprecedented threats due to human activities, such as deforestation, illegal logging, and climate change. It is therefore crucial that we make continued efforts in forest conservation to ensure the sustainability of these invaluable ecosystems.

Conservation biology plays a vital role in understanding and addressing the challenges faced by forests. This interdisciplinary field combines biology, ecology, and environmental science to study the natural world and develop strategies for its protection. By studying the ecological processes that occur within forests, conservation biologists can identify the key species and habitats that need our attention and develop conservation plans to safeguard their future.

One of the main reasons why continued efforts in forest conservation are so important is the role forests play in combating climate change. Forests act as carbon sinks, absorbing large amounts of carbon dioxide from the atmosphere and storing it in their biomass. By preserving and restoring forests, we can help reduce the amount of greenhouse gases in the atmosphere and mitigate the effects of climate change.

Furthermore, forests are home to countless species, many of which are found nowhere else on Earth. These unique ecosystems provide habitat for numerous plants, animals, and microorganisms, contributing to biodiversity conservation. By protecting forests, we are not only saving individual species but also preserving entire ecosystems and the ecological services they provide.

Forest conservation also has significant socio-economic benefits. Many communities around the world rely on forests for their livelihoods, through activities such as sustainable logging, ecotourism, and the collection of non-timber forest products. By promoting sustainable forest management practices, we can ensure the well-being of these communities and create economic opportunities that are compatible with long-term forest conservation goals.

In conclusion, the importance of continued efforts in forest conservation cannot be overstated. Forests are critical for the well-being of our planet and its inhabitants. By understanding the ecological processes within forests and implementing sustainable management practices, we can protect these invaluable ecosystems and ensure their survival for future generations. Whether you are a conservation biologist or an individual passionate about the environment, it is up to every one of us to take action and contribute to the conservation of our world's forests.

Encouraging Everyone to Engage in Forest Conservation

In today's fast-paced world, where our attention is often drawn to urban landscapes and technological advancements, it is crucial that we do not forget the importance of our forests. Forests are not only home to an incredible diversity of flora and fauna but also play a vital role in regulating our climate, purifying the air we breathe, and providing us with essential resources. However, the responsibility of conserving these invaluable ecosystems does not solely lie in the hands of scientists and policymakers. It is a task that concerns each and every one of us.

Conservation biology is a field that aims to understand and protect the natural world, and engaging everyone in this endeavor is essential. By emphasizing the importance of forest conservation to all individuals, we can create a global movement that will have a lasting impact on our planet. Whether you are a student, a teacher, a parent, or a professional in any field, there are numerous ways you can contribute to the conservation of forests.

Educating ourselves and others about the value of forests is a crucial first step. By learning about the intricate web of life that exists within these ecosystems, we can develop a deeper appreciation for their significance. This knowledge can then be shared with our friends, families, and communities, creating a ripple effect of awareness and understanding.

Taking action is equally important. Even small changes in our daily lives can make a big difference. Planting native trees in our gardens, reducing our consumption of paper products, and supporting sustainable forestry practices are just a few examples of how we can actively contribute to forest conservation.

Engaging in citizen science initiatives is another powerful way to get involved. By participating in projects that collect data on forest biodiversity or monitor the health of local ecosystems, we can provide valuable information for scientists and contribute to their conservation efforts.

Finally, supporting organizations and initiatives that focus on forest conservation is instrumental in driving change at a larger scale. Donating to reputable nonprofits, volunteering for local conservation projects, or joining advocacy groups can all have a significant impact on the protection of forests worldwide.

In conclusion, forest conservation is not limited to a specific group of individuals. It is a responsibility that falls upon us all. By educating ourselves, taking action, participating in citizen science, and supporting relevant organizations, we can collectively make a difference. The future of our forests depends on our willingness to engage, and together, we can ensure the preservation of these incredible ecosystems for generations to come.